Practical
PROPAGATION

Alan Toogood

The Crowood Press

First published in 1992 by
The Crowood Press Ltd
Ramsbury, Marlborough
Wiltshire SN8 2HR

British Library Cataloguing in Publication Data

A catalogue record for this book is available from the
British Library.

ISBN 1 85223 631 0

Acknowledgements

Line drawings by Claire Upsdale-Jones.
All photographs by Sue Atkinson.
Thanks to: Steve Bradley of Merristwood College of
Agriculture and the Department of Arboriculture,
Worplesdon, Surrey; Mr Simpson, Kennedy's Garden
Centre, Twyford, Bucks; Mr Tim Dorset, Bourne End
Nurseries, Bourne End, Bucks; The Willows Garden Centre,
Windsor, Bucks; William Wood and Sons, The Bishop
Centre, Bath Road, Taplow, Bucks.

Typeset by Chippendale Type Ltd, Otley, West Yorkshire
Printed and bound in Great Britain by
BPCC Hazell Books, Aylesbury

CONTENTS

INTRODUCTION

For thousands of years people have been fascinated by plant propagation. Ancient civilizations, including the Romans, practised various techniques, some of which are still in use today. Propagation has never lost its appeal. Indeed, plants are being increased on a large scale today by home gardeners.

With plants being freely available from garden centres, nurseries and even high street chain stores, why should we want to propagate? There are various reasons. Firstly, it is a fascinating and enjoyable aspect of gardening and some people do it for the sheer pleasure of producing new plants. Secondly, it is an economical way of obtaining more plants for your garden, greenhouse and home. Admittedly you will be producing more of those you already have, but this is invariably a good thing as many plants are best grown in groups of each cultivar or species for greater impact. Thirdly, surplus plants can be given away to gardening friends, neighbours and relations. One can even start up a plant-swapping scheme among friends and neighbours. In

this way, all concerned eventually obtain a wide range of plants.

Raising plants from seeds bought from seedsmen will greatly increase the range of plants in your garden for a very modest outlay. Seeds are still the cheapest way of obtaining new plants, even though some people do complain about the prices on the packets!

Some aspects of plant propagation are easier than others for the home gardener. I have organized this guide to begin with easier techniques, gradually progressing to more skilled operations — from division of plants, through layering, plantlets, cuttings of all kinds, seed sowing (outdoors and under cover), to budding and grafting.

The Seasonal Tasks table will help you to carry out plant propagation at the optimum time throughout the year, timing being one of the keys to success (*see* pages 60–61).

All the major methods of plant propagation are well illustrated with drawings, as words alone are inadequate to describe this fascinating art.

The easiest method of propagation is division, suitable for many clump-forming plants or those which form offsets, including such exotic kinds as bromeliads.

1 • DIVISION

Division is the easiest method of plant propagation. It simply involves splitting a plant, or a group of plants, into a number of portions. Only clump-forming and carpeting plants, and those which produce offsets, can be divided.

Hardy Perennial Plants

Included under this heading are hardy herbaceous and evergreen perennials, alpines (or rock plants), and aquatics (or water plants).

Hardy herbaceous and evergreen perennials, perhaps better known as border plants, include achilleas, asters (Michaelmas daisies), bergenias, *campanula* (bellflower), chrysanthemums, *geranium* (crane's-bill), *hemerocallis* (day lily), hostas (plantain lily), *phlox*, *solidago* (golden rod), ferns and ornamental grasses.

Border plants should be lifted and divided every three or four years to ensure that they grow and flower better. It will also provide you with more plants. Division can take place between early and mid-spring. Mid- to late autumn is suitable only if the soil is extremely well drained. Early-flowering kinds, like *doronicum* (leopard's

Clump-forming border plants such as asters (Michaelmas daisies) should be lifted and divided every three to four years to keep them young and vigorous.

Bearded irises can be divided immediately after they have flowered, each division consisting of a portion of rhizome with roots and a reduced fan of leaves.

bane), *epimedium* (barrenwort), *pulmonaria* (lungwort) and bearded irises, can be divided immediately after flowering.

The technique is clearly shown in the accompanying panel. Note that the method is slightly different for bearded irises when each division consists of a portion of rhizome with roots and a fan of leaves.

Clump-forming and carpeting alpines, like saxifrages, are divided in the same way,

Clump-forming alpines such as saxifrages are divided in the same way as hardy border perennials, the best time being during early spring.

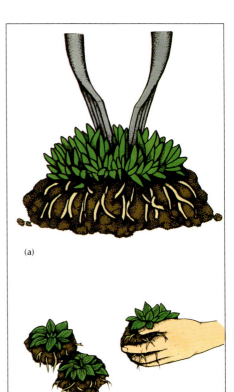

(a)

(b)

Dividing a Hardy Herbaceous Perennial
Lift plant with a garden or hand fork, depending on size, and shake the roots free of soil. A good way to divide large clumps is to insert two forks, back to back, through the centre and prize the handles apart (a). Repeat to divide further. Smaller plants can be split by hand – just pull them apart.
• **Replanting**   Each division should consist of growth buds plus a good proportion of fibrous roots. Herbaceous divisions can be about hand sized (b). Always discard the centre part of a plant as this is old and declining in vigour, saving only the young outer parts. Replant divisions immediately and never allow them to dry out.

Rooted offsets of sempervivums
(houseleeks) and similar alpines can be
carefully removed in early spring and
replanted elsewhere.

*Water-lilies are propagated by division in
mid-spring to early summer, the tubers
being cut into pieces complete with
growth buds and roots.*

An aspidistra *badly in need of division.
Many other clump-forming greenhouse
and houseplants can also be propagated
by this method in the spring.*

during early spring. Some alpines such as
sempervivum (houseleek) produce offsets.
Rooted offsets can be carefully removed
and replanted elsewhere.

Aquatics are divided similarly, during the
period mid-spring to early summer. With
water-lilies, the tubers can be cut into pieces,
complete with some growth buds and roots.

Greenhouse Plants and Houseplants

Greenhouse plants and houseplants which
form clumps can also be divided, the technique being virtually the same as for border
plants. The best time to do this job is early
to mid-spring. Examples include *aspidistra*
(cast-iron plant), bromeliads, numerous
cacti and succulents, *chlorophytum* (spider

This bromeliad has several well-developed offsets. These can be removed, with some roots if possible, and potted into small containers to form new plants.

plant), *sansevieria* (mother-in-law's tongue) and ferns.

Remove the plant from its pot, tease soil away from the roots and then pull the plant apart into a number of smaller portions. If this is difficult to do, use a knife. Pot into suitably sized containers.

Some bromeliads (like the urn plant) produce offsets. When these are around 6in (15cm) high, carefully remove (with some roots if possible), pot into small containers and encourage to establish in a heated propagator.

Bulbs

This category includes true bulbs like daffodils and tulips, as well as corms such as

Congested clumps of bulbs can be separated when dormant. The largest daffodil bulbs will have multiple growing points (left); and the smallest, single 'noses' (right).

Mature gladiolus corms produce many offsets, known as cormlets, which can be removed and grown on to flowering size in a nursery bed.

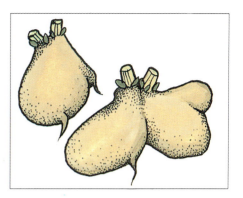

Dahlia divisions must have some growth buds at the base of the old stem, plus a tuber, in order to grow.

crocuses and gladioli, and tubers, an example of which being dahlias.

Congested clumps of bulbs and corms can be dug up when the leaves have died down (for most, late spring or summer), and separated into single bulbs. These should be dried off and stored until planting time, which is early to mid-autumn for most. *Galanthus* (snowdrop) are best lifted, divided and replanted immediately after flowering.

Some bulbs, like daffodils, freely produce bulblets or offsets. These can be detached if desired and replanted in a nursery bed to grow on to flowering size.

Cormous plants like crocuses and gladioli produce tiny offsets, known as cormlets, around themselves. These can be removed from dormant corms and planted in a nursery bed, in the same way as bulblets, to grow on to flowering size. However, cormlets of the tender gladioli must be stored in a tray of dry peat in a cool but frost-proof place over winter and planted in mid-spring, lifting them again in autumn for winter storage.

Large dormant clumps of dahlias can be divided in mid-spring, just prior to planting in the garden. Each division must consist of

Daffodils and similar bulbs freely produce bulblets or offsets (this is how they eventually form clumps, as here), which can be detached and planted elsewhere.

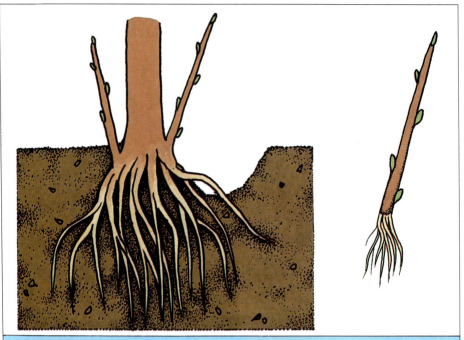

Shrubs from Suckers
Some shrubs produce suckers from the roots which can be used for propagation. These are simply shoots with roots attached.
· **Suckering Shrubs** Examples include *rhus* (stag's horn sumach) shown here, *symphoricarpos* (snowberry), raspberries and clump-forming shrubby *cornus* (dogwoods).
· **Removing Suckers** Carry out sucker removal when the shrubs are dormant, between late autumn and early spring. Remove without disturbing the parent plant by carefully digging away soil until the sucker's point of origin is found. Then cut it off there, complete with roots. Replant suckers immediately, elsewhere, as they should not be allowed to dry out. Remember never to remove suckers from budded or grafted shrubs or trees for the purpose of propagation as these will originate from the rootstock (*see* Chapters 8 and 9). Rootstock suckers do not make good garden plants and, therefore, should be removed and destroyed.

Rhus typhina *suckers*.

some growth buds plus at least one tuber. Some clumps may be difficult to divide, necessitating the use of a knife. Remember to lift dahlias in autumn and to store them in a cool but frost-proof place over winter, as they are tender perennials.

2 • LAYERING

Layering is another very easy method of propagation. It is used for shrubs, trees, climbers, woody house-plants and certain berried fruits and carnations. It involves encouraging a shoot or stem to produce roots while it is still attached to the parent plant. When rooted, the shoot is severed and planted, when it grows into a new plant. These new plants can be grown on in pots or simply set out in flower beds depending on the type. This method entails little work and can give excellent results.

Simple Layering

This technique is ideal for propagating shrubs which are difficult from cuttings, such as rhododendrons, camellias, magnolias, *hamamelis* (witch hazel) and *pieris*. However, you can try layering any shrub provided shoots or branches can easily be brought down to the ground.

The best time to layer is between mid-spring and late summer. You must choose young shoots produced in the current or

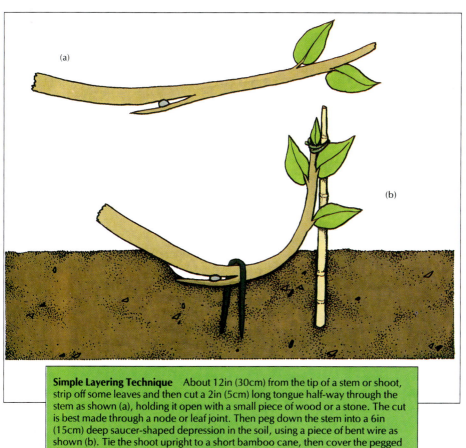

(a)

(b)

Simple Layering Technique About 12in (30cm) from the tip of a stem or shoot, strip off some leaves and then cut a 2in (5cm) long tongue half-way through the stem as shown (a), holding it open with a small piece of wood or a stone. The cut is best made through a node or leaf joint. Then peg down the stem into a 6in (15cm) deep saucer-shaped depression in the soil, using a piece of bent wire as shown (b). Tie the shoot upright to a short bamboo cane, then cover the pegged area of stem with 6in (15cm) of soil, firming it lightly.

previous season as old woody growth will not root.

Layers should be rooted in well-prepared soil. Dig the ground about 12in (30cm) deep, at the same time mixing into it some peat and coarse horticultural sand. Break down the soil finely.

Preparation and pegging down of the shoot or stem are described in the accompanying panel. Keep layered stems moist and they will form a good root system within eighteen to twenty-four months, depending on the type of shrub. In autumn or early spring lift with a fork, sever from the parent plant and immediately replant elsewhere.

Air Layering

Air layering is a variation on simple layering. This particular method is used for trees and shrubs whose branches cannot be brought into contact with the soil.

The shoot is prepared as for simple layering, but the tongue is held open with a wad of moist sphagnum moss. The prepared area of stem is then wrapped in additional moss, covered with a 'bandage' of clear polythene sheeting, and sealed with waterproof tape.

Roots may form within twelve to twenty-four months, at which point the polythene is removed and the rooted stem is cut away

With air layering, the 'tongue' is packed and wrapped with sphagnum moss (left) which is held in place with a bandage of polythene (right).

Air Layering a Houseplant   Woody houseplants like ficus (rubber plants and figs), codiaeums (crotons) and philodendrons can be air layered in a warm room during summer. A stem is prepared about 12in (30cm) from the top (for the basic technique *see* pages 11–12). The cut to form the tongue should be made in an upward direction and then dusted with hormone rooting powder to speed rooting. The stem should root within a matter of weeks in warm conditions. It is then removed as for outdoor air layers but potted into a suitably sized container. If desired, the parent plant can be cut back, when it should make new bushy growth.

for planting in the garden. There is no need to remove the moss.

Serpentine Layering

This method is recommended for climbing plants like *jasminum* (jasmine), *clematis*, *lonicera* (honeysuckle) and wisterias. It is a variation on simple layering, the difference being that instead of rooting a stem in just one place, a long young stem is chosen and pegged down in a number of places along its length. Thus, each stem will produce a number of new plants. Many climbers root more quickly than shrubs and it is quite

Tip Layering Blackberries, loganberries and other similar long-stemmed berrying fruits can be propagated by tip layering in the period mid- to late summer. Use young stems produced in the current year. Prepare the soil as described under Simple Layering, page 12, unless rooting in a pot. The tip of the stem is buried about 3in (8cm) deep in the ground, or in a flowerpot containing a mixture of peat and sand. The pot can be sunk into the ground. By late autumn the tip will have rooted, when it can be severed, lifted and planted elsewhere.

possible that some may be sufficiently well rooted to allow lifting within twelve months.

Layering Carnations

It comes as a surprise to many people that border carnations can be easily propagated

Clematis and numerous other climbing plants are easily propagated by serpentine layering, whereby a single stem can produce a number of new plants.

by layering. For many people this may prove to be easier than taking cuttings. Border pinks can also be layered, but as they have much thinner stems you may find them more difficult. If so, you should resort to cuttings (*see* Pipings, page 38). It is essential to use the current year's shoots before they start to become hard and turn woody. Layering can be carried out after flowering, probably in late summer.

Shallowly loosen the soil around the parent plant with a hand fork and then spread a 2in (5cm) deep layer of soil-based

Although layering is more usually associated with shrubs and other woody plants, it can also be used to propagate border carnations during late summer.

potting compost around it. Use shoots which have not flowered for layering, removing the leaves from an area near their base where they will be pegged down. Then cut a 1½in (3.5cm) long tongue in the stem, half-way through it and through a node (*see* Simple Layering, page 11).

This prepared area of stem is then pegged down into the layer of compost with a wire peg, making sure that the tongue remains open, and covering it with a 2in (5cm) deep layer of potting compost. A number of stems can be layered around the parent plant.

If the layered stems are kept moist, they should form good root systems within two months. When they do, cut the stems from the parent plant, and a few days later lift the rooted layers and replant them elsewhere. They will flower the following year.

Carnations and pinks should be replaced with young plants on a regular basis as they are short-lived perennials and soon start to deteriorate. Propagation ought to take place every two or three years.

3 • PLANTLETS

Some plants produce small plantlets on their leaves or stems and these can be used for propagation. For these plants it is a natural form of increase.

Plantlets on Leaves

Most plants which produce plantlets on their leaves are tender and, therefore, grown indoors or in a greenhouse. A good and popular example is *Kalanchoe daigremontiana* (Mexican hat plant), a succulent that produces little plantlets along the edges of its leaves. Another succulent species, *K. tubiflora*, carries plantlets on the ends of long cylindrical leaves. Eventually the plantlets drop off and root into the soil.

Little bulb-like organs which develop into tiny plants grow on the leaves of the tender fern Asplenium bulbiferum *(spleenwort). These can be removed and planted in containers to root.*

Kalanchoe daigremontiana *(Mexican hat plant) produces little plantlets along the edges of its leaves and these can be picked off and rooted in pots or trays.*

If you allow this to happen, the plantlets could fall anywhere and be lost. It is, therefore, better to pick them off, plant shallowly in a pot, or tray, of potting compost and root them in a propagator with a temperature of 15.5°C (59.9°F).

The tender fern *Asplenium bulbiferum* (spleenwort) produces little bulb-like organs on its fronds (leaves) and these grow into tiny plants. In nature, when they come in contact with the soil (either by dropping off or when the frond dies), they take root. You can speed this up by removing plantlets when well developed and planting them in potting compost. They will root in moderate heat.

Plantlets on Stems

Some plants produce plantlets on long stems. One of the best-known and most popular examples is the strawberry. The stems produced by strawberry plants in the summer are correctly known as runners. As they grow, they produce buds which then

Tolmiea menziesii 'Pick-a-Back Plant' This perennial plant, also known as youth-on-age, is hardy and can be grown outdoors, although many people grow it as a houseplant. *Characteristics* The somewhat ivy-shaped leaves are evergreen or semi-evergreen and when mature carry plantlets where they join the stalk.

Propagation Leaves with well-developed plantlets can be pegged down into the soil while still attached to the parent plant, as shown here. During spring or summer they will root within a matter of weeks and can then be removed and potted separately. *Alternative method* In spring or summer remove the leaves with well-developed plantlets, complete with leaf stalks, and insert in pots or trays of potting compost so that the base of each plantlet is in close contact with the compost. In moderate heat, rooting will be quick.

root into the soil and develop into new plants.

Propagating Strawberries

Generally lots of runners are produced and they should be removed if not required for propagation. However, bear in mind that strawberry plants need replacing with young ones every three years so it is advisable to propagate on a regular basis for best results.

To encourage the plantlets to root quickly into the soil they are pegged down so that they are in close contact with the ground. This is best done in early or mid-summer.

The first plantlet produced on a runner (the one nearest the parent plant) is the best one for pegging down to produce a new plant. Remove runners produced beyond the first plantlets.

There is a limit to the number of plantlets that each parent plant should be allowed to produce – five or six is the maximum.

The first plantlet produced on a strawberry runner is the best one to peg down to produce a new plant. Surplus runners should be removed.

Remove all surplus runners before they become too well developed.

Pegging down plantlets is simple – insert a wire peg over the runner, just behind a plantlet, and push it down into the soil so that the base of the plantlet is in close contact with the ground.

The plantlets can be rooted in the garden soil, which should first be shallowly loosened with a hand trowel, taking care to avoid disturbing the parent plant, but it would be better to root them in 3in (8cm) pots of potting compost, sunk to their rims in the soil. This will ensure little or no root disturbance when eventually the young plants are installed elsewhere.

It is vital to keep the soil moist during the rooting period as rooting will not occur, or will be poor, if the ground is allowed to dry out.

The plantlets will root quickly and by late summer or early autumn they should be severed from the parent plant, lifted and planted in a new strawberry bed, or in strawberry barrels or pots. The young plants must be kept moist after planting so that they can establish quickly.

Young strawberries planted in late summer or early autumn will start to produce fruit the following year.

Other Plants

A houseplant that produces little plantlets

Variegated cultivar of Saxifraga stolonifera *which, like its parent, produces little plantlets on long stems. When these come into contact with soil, they form roots.*

Chlorophytum 'Spider Plant'   Chloro-phytums, of which the green-and-white striped *C. comosum* 'Variegatum' is the best known, are houseplants which produce plantlets on the ends of long flower stems, following the flowers. These attractive plants are best grown in elevated positions so that the plantlets can hang down.
Propagation   This can be achieved during spring or summer by pegging down well-developed plantlets into 3in (8cm) pots placed alongside the parent, as shown here. Fill the pots with potting compost and use wire pegs.
Aftercare   The plantlets should root into the compost within six weeks if kept in a warm room, when they can be severed from the parent.

on long stems is *Saxifraga stolonifera* (also known as *S. sarmentosa*). You may know it under its popular name of mother of thousands. It is an evergreen perennial which produces masses of thin runners (rather like those of strawberries) carrying plantlets at their ends. When these come into contact with the soil, they root into it. The best way to appreciate this attractive plant is to elevate it so that the runners can hang down like a curtain.

To propagate, well-developed plantlets can be pegged down individually into 3in (8cm) pots of potting compost placed along-side the parent plant, in the same way as strawberry runners. Summer is the best time for this.

After about six weeks the plantlets should be well rooted in their pots; when they can be severed from their parent. They can then be left to grow on into strong, vigorous new plants.

4 • STEM CUTTINGS

Stem cuttings are shoots (generally side shoots) which are removed from plants and encouraged to form roots of their own. A wide range of hardy and tender plants can be propagated from stem cuttings but, while many subjects root easily, others (not mentioned in this book) can present difficulties to the amateur. Varying conditions are needed for rooting stem cuttings – soft cuttings generally need artificial heat while harder or more woody ones will often root in cooler conditions.

Softwood Cuttings

Side shoots formed in the current year are used for softwood cuttings. They are prepared while the shoots are soft, generally in the period mid-spring to early summer.

Hardy plants, including a wide range of shrubs, climbers, perennials and alpines, can be propagated from softwood cuttings, plus many tender or greenhouse plants including *coleus* (flame nettle), fuchsias, *impatiens* (busy lizzie) and *tradescantia* (wandering Jew). A good rule is that if a plant produces plenty of soft young side shoots, it is worth trying to propagate.

Try to prevent cuttings from wilting between removal from the parent plant and rooting. Collect them in a polythene bag and work in cool, shady conditions.

Preparation and insertion of cuttings is shown in the accompanying panel. As soon as they have been inserted, transfer them to a heated propagator and keep it closed to maintain high humidity, but ventilate for an hour or so several times a week. A good temperature is 15.5°C (59.9°F).

A heated greenhouse is an ideal place to root the softer stem cuttings.

(a)

Preparing and Inserting Softwood Cuttings
Cuttings should be up to 3in (8cm) in length depending on the subject (some may be much shorter, like those of alpines) and prepared from the tops of side shoots. Use a sharp knife or razor blade and make the basal cut immediately below a node, or leaf joint, and then cut off the bottom leaves so that the lower third of the cutting is free from foliage (a). Dip the lower ⅕in (6mm) in hormone rooting powder formulated for softwood cuttings.
• **Containers and compost** Cuttings can be rooted in plastic pots, or in larger quantities in seed trays. Use a home-made compost consisting of equal parts moist sphagnum peat and coarse horticultural sand.
• **Insertion** Make a hole for each cutting with a pencil and insert up to the lower leaves, checking that the base is in close contact with the bottom of the hole (b). Firm the compost around it. After inserting cuttings, water them in.

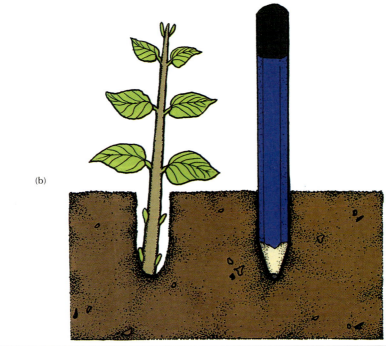

(b)

Soft Basal Cuttings   Hardy and tender perennials which produce few if any side shoots like delphiniums, dahlias, chrysanthemums, lupins, gypsophilas (chalk plant) and campanulas (bell-flowers) can be increased from soft basal cuttings in spring. When plants have started into growth, remove some shoots when about 2in (5cm) high, cutting them off very close to their base. Remove lower leaves and dip the bases of the cuttings in hormone rooting powder. Insertion and rooting conditions are the same as for ordinary softwood cuttings.

Alternatively, enclose the container in a clear polythene bag, supported above the cuttings with a few short split canes, and place on a window sill in a warm room.

Tips of cuttings starting into growth is an indication of rooting. They can then be lifted carefully and potted into 3–3½in (8–9cm) containers, using potting compost. Young hardy plants will need hardening (slowly acclimatized to outdoor conditions) in a garden frame before planting outside.

Keep a regular check on how they progress before determining their readiness for outdoor planting.

Semi-Ripe Cuttings

Current year's shoots are also used for semi-ripe cuttings. These are taken later in the year, between mid-summer and mid-autumn, hence they are more woody or 'riper' than softwood cuttings. The bases are firm and woody but the tops are still soft. These are generally easier to root than softwood cuttings and many do not need artificial heat, being rooted in a cold greenhouse or garden frame.

A very wide range of plants can be increased from semi-ripe cuttings: hardy

Preparing and Inserting Semi-Ripe Cuttings These are very similar to softwood cuttings except the lower part is more woody. The average length for shrubs and conifers is 4–6in (10–15cm). Dip bases in hormone rooting powder. Use plastic pots or deep seed trays, filling them with a mixture of equal parts moist sphagnum peat and coarse horticultural sand. Alternatively, root directly in the soil in a garden frame, but first mix plenty of peat and coarse sand into it. The insertion technique is the same as for softwood cuttings.

shrubs like *berberis* (barberries), callunas and ericas (heathers), cotoneasters, escallonias, hebes (shrubby veronicas), hydrangeas, lavenders, potentillas (shrubby cinquefoils), *pyracantha* (firethorns) and viburnums; conifers such as *chamaecyparis* (false cypress) and junipers; climbers, for example, *jasminum* (jasmines); half-hardy perennials like pelargoniums; and greenhouse plants or house-plants, such as abutilons, bougainvilleas (paper flowers), *citrus* (oranges, lemons, grapefruits, etc.), fuchsias, *justicia* (shrimp plant) and *plumbago*.

Variations

The panel illustrates the preparation of a normal semi-ripe cutting, but there are a few variations. If you are propagating *berberis* (barberries) with thin shoots, use mallet cuttings as shown here. These consist of the current year's shoots removed with about ⅕in (6mm) of the main woody stem.

Semi-ripe cuttings of berberis with thin shoots will stand a better chance of rooting if they have a piece of woody stem attached.

These stand more chance of rooting than thin shoots.

Conifer cuttings are like normal cuttings, but one must ensure that their bases are really well ripened and have a minimum of ⅖in (1cm) of brown wood to allow successful rooting.

Cuttings of ericas (left), and callunas (right) – heathers – are obtained by pulling off side shoots with a piece of woody tissue attached.

Semi-ripe cuttings of conifers like chamaecyparis *must have well-ripened bases to ensure successful rooting. A minimum of ⅖in (1cm) of brown wood is recommended.*

The smallest semi-ripe cuttings are those of callunas and ericas (heathers). These are only 1¼–2in (3–5cm) in length. They are obtained by pulling off side shoots so that each has a piece of woody tissue attached. This is known as a 'heel'. Strip off the lower leaves between finger and thumb, except for calluna cuttings, and pinch out tips.

Rooting Conditions

It has already been said that many semi-ripe

Pelargoniums produce numerous side shoots which can be used as semi-ripe cuttings, rooting them during late summer. They do not need artificial heat to root.

Heathers, such as callunas, are propagated from tiny semi-ripe cuttings, no more than 2in (5cm) in length. They are obtained by pulling off side shoots with a heel.

cuttings can be rooted without artificial heat, in a cold greenhouse or garden frame; this certainly applies to hardy plants. However, bear in mind that some may take many months to root, especially those taken late in the year. Cuttings taken in late summer and autumn will not be well rooted until the following spring. Those taken earlier will have good root systems by autumn.

However, it should be said that rooting will be quicker if they are placed in a heated propagator. Cuttings of tender plants should definitely be rooted with heat. A bottom temperature of 15.5°C (59.9°F) will be sufficient for all subjects and there is no need to provide very high temperatures; this may, in fact, adversely affect rooting. (*See* Softwood Cuttings, page 21, for further details of rooting conditions.)

When cuttings have rooted, they can be lifted and planted in pots of suitable size. Hardy plants raised in heat will need to be hardened in a garden frame prior to planting out in the garden. They should remain in the frame over a period of several weeks and subjected to gradual ventilation. Cuttings which have been rooted in the soil in a garden frame can be lifted and planted straight away in the garden if desired, instead of potting them.

Hardwood Cuttings

Hardwood cuttings are the easiest of the stem cuttings. They are simple to prepare, need no artificial heat and are almost guaranteed to root. The current year's

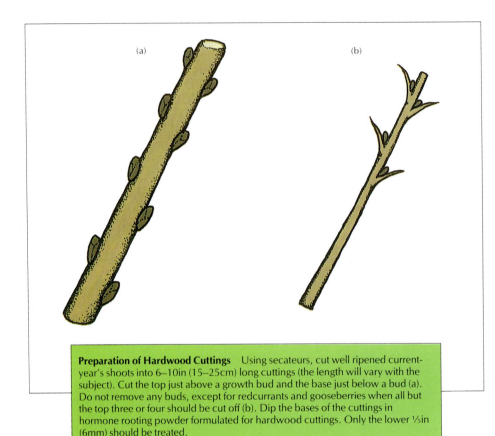

(a)   (b)

Preparation of Hardwood Cuttings Using secateurs, cut well ripened current-year's shoots into 6–10in (15–25cm) long cuttings (the length will vary with the subject). Cut the top just above a growth bud and the base just below a bud (a). Do not remove any buds, except for redcurrants and gooseberries when all but the top three or four should be cut off (b). Dip the bases of the cuttings in hormone rooting powder formulated for hardwood cuttings. Only the lower ⅕in (6mm) should be treated.

shoots are used once again, but this time they are fully ripe – hard and woody. They are taken late in the year, between late autumn and early winter, when the plants are dormant.

Various hardy plants can be propagated from this type of cutting, including the shrubs *buddleia* (butterfly bush), *cornus* (shrubby dogwoods), *deutzia, forsythia* (golden bell), *ligustrum* (privet), *Lonicera nitida* (Chinese honeysuckle), *philadelphus* (mock orange), *ribes* (flowering currant), *salix* (willows), spiraeas, viburnums and *weigela*; and the soft fruits gooseberries, blackcurrants, redcurrants and white currants.

Rooting the Cuttings

Hardwood cuttings can be rooted either in the open ground or in a garden frame, depending on the subject. Those which will root successfully out of doors are *ligustrum* (privet), *Lonicera nitida* (Chinese honeysuckle), *salix* (willows), gooseberries and blackcurrants, redcurrants and white currants. All the other subjects mentioned earlier would be better inserted in a garden frame. This will protect them from icy drying winds and wet soil conditions over the winter.

For outdoor rooting choose a warm sunny spot sheltered from cold winds and with well-drained soil. To improve drainage dig in plenty of grit or coarse horticultural sand. A garden frame should also be located in a sunny sheltered position. Again, dig in plenty of grit or sand if the soil is not too well drained.

The cuttings are inserted one-half to two-thirds of their length into the soil. Make a V-shaped trench with a garden spade and then place in the cuttings, ensuring that they are upright, and spacing them 4–6in (10–15cm) apart. Firm the soil around them well by treading with your heels. If the ground is soft, you may be able to push the cuttings into the soil; this method is quicker and easier than removing a trench.

Aftercare

Outdoors, cuttings may be lifted by frost during the winter. If this happens, it is essential to refirm them immediately the ground thaws out.

The cover or lid of a garden frame should be closed as soon as any cuttings have been inserted and kept closed throughout the winter, although you should check the cuttings occasionally to make sure that they are not drying out and to remove any dead

Hardwood cuttings are inserted to one-half to two-thirds of their length in the soil. This is easily achieved if a V-shaped trench is made for them.

Shrubby dogwoods or cornus *noted for their coloured bark are easily propagated from hardwood cuttings taken between late autumn and early winter and rooted in a garden frame.*

ones. The frame can be fully opened in mid- to late spring.

The cuttings should be watered in the following spring and summer if the soil starts to dry out. Lack of water will prevent them from rooting. Cuttings in outdoor or frame beds will need to be weeded during the spring and summer.

By autumn of the following year, one year from insertion, the cuttings will be well rooted and can be lifted carefully and planted in another part of the garden.

Many people think that hardwood cuttings will be well rooted by the spring of the following year, but this is not the case. Certainly they will have produced leaves and shoots (this is what makes people think they have rooted), but it takes longer for roots to form. You will have to be patient – leave them in place for a year.

During the course of the rooting year some subjects will put on quite a lot of top growth, especially vigorous shrubs like *buddleia* (butterfly bush) and *salix* (willows). With these you will have reasonably large young plants for setting out in beds and borders.

5 • SOME UNUSUAL CUTTINGS

Cuttings can be made from parts of plants other than stems, including roots and leaves. The techniques involved embrace some of the most fascinating aspects of plant propagation and, what is more, most are easy and quite within the range of home gardeners.

Root Cuttings

Numerous hardy plants can be propagated from root cuttings, including the shrubs *Aralia elata* (Japanese angelica tree) and *Rhus typhina* (stag's horn sumach); and the perennials *anchusa* (alkanet), *Anemone hupenhensis japonica* and *A. hybrida* (Japanese anemones), *eryngium* (sea holly), gaillardias (blanket flowers), geraniums (crane's-bill), *Papaver orientale* (oriental poppy), *Phlox paniculata* (border phlox) and *Primula denticulata* (drumstick primrose).

The time to take root cuttings is when plants are dormant, during early and midwinter. The method of collection will

The 'architectural' shrub *Aralia elata* (Japanese angelica tree) can be propagated from root cuttings taken during the dormant season.

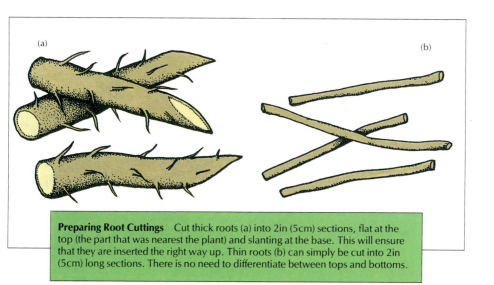

(a)

(b)

Preparing Root Cuttings   Cut thick roots (a) into 2in (5cm) sections, flat at the top (the part that was nearest the plant) and slanting at the base. This will ensure that they are inserted the right way up. Thin roots (b) can simply be cut into 2in (5cm) long sections. There is no need to differentiate between tops and bottoms.

depend on the size of the plant. With a large shrub, such as *Rhus typhina* (stag's horn sumach), you will have to scrape the soil away until some young roots are exposed. In the case of a hardy perennial, like *Phlox paniculata* (border phlox), the whole plant can be dug up to have some roots removed, then replanted.

Most of the plants mentioned have thick roots. Choose those about the thickness of a pencil for cuttings. *Phlox paniculata* (border phlox) has thin roots and any of these, except the hair-like ones, can be used. Plants should never have all their roots removed – take only a few from each specimen, cutting them off cleanly with a sharp knife or secateurs.

Rooting the Cuttings

Preparation of cuttings is described in the accompanying panel. They can be inserted in deep pots or seed trays filled with a mixture of equal parts moist sphagnum peat and coarse horticultural sand.

Thick root cuttings are inserted 2in (5cm) apart in the compost simply by pushing them in until the tops are just below the surface. They should be covered with a thin layer of compost. Thin cuttings are simply laid 1in (2.5cm) apart on the surface of the compost and covered with a ⅖in (1cm) layer of compost which should be firmed lightly.

Cuttings should be well watered after insertion, allowed to drain then placed in a garden frame or cold greenhouse. Keep the frame lid or cover on over winter and remove it in mid-spring.

Over winter check that the compost remains slightly moist, but not wet or the cuttings may rot. The cuttings should be lifted as soon as they have rooted, which will be the following spring or summer. You will notice, though, that they produce top growth (shoots and leaves) in the spring; this does not mean they have formed roots, which generally follow later.

Rooted cuttings can either be potted and placed in a garden frame to establish, or they can be planted direct in the garden.

Leaf Cuttings

Numerous tender plants and a few hardy

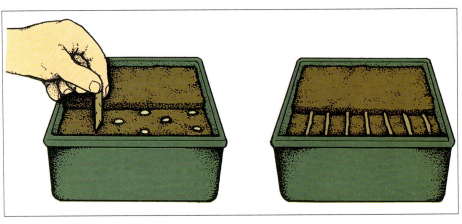

Thick root cuttings are inserted by pushing them down into the compost (left), while thin ones are laid flat and then covered (right).

kinds can be raised from leaf cuttings during spring or summer — from either whole leaves or small sections of leaves. This is a fascinating method of propagation, and, with most plants, success is assured given the right rooting conditions.

Whole Leaves

Tender plants that can be raised from whole leaves include *peperomia* (pepper elder), *saintpaulia* (African violet), *Sinningia* (gloxinias), echeverias, crassulas, aloes, sedums, *streptocarpus* (Cape primrose) and various begonias like *B. rex* and *B. masoniana*. Several hardy rock-plants can be raised from whole leaves: lewisias, ramondas and haberleas.

For plants with small leaves (all except the begonias, *sinningia* and *streptocarpus*), simply cut off a few mature leaves complete with leaf stalks and dip the bases in hormone rooting powder. Then insert them in pots of cutting compost (equal parts moist

Entire leaves of peperomia (pepper elders), complete with stalks, can be rooted in a warm propagator during spring or summer to form new plants.

sphagnum peat and coarse horticultural sand) so that the leaf stalks are completely buried. Lightly firm them in and water well.

Mature leaves, complete with stalks, provide a means of increasing saintpaulia *(African violets). Root them during the summer in a heated propagator.*

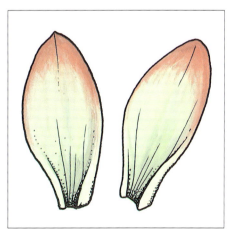

Numerous succulent plants, such as echeveria, *can be propagated from whole leaves during the summer. Insert them only shallowly in the cutting compost.*

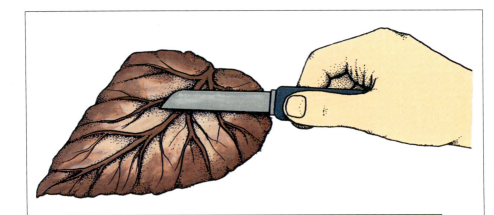

Propagating Begonias

Large-leaved begonias, especially *B. rex* and *B. masoniana*, are propagated from whole leaves.

• **Technique**   Remove a mature leaf, cut off the stalk, turn the leaf upside-down and then cut through the main veins in a number of places with a sharp knife. Lay the leaf, the right way up, on the surface of cutting compost (*see* Whole Leaves, page 33) and weight it down with a few small stones or pebbles to make sure that the cut veins are in close contact with the compost.

• **Rooting**   For rooting conditions *see* Whole Leaves, page 33. Little plantlets will appear where the veins have been cut and these can be removed and potted individually.

Large whole leaves of the popular iron-cross begonia, Begonia masoniana, *can be used for propagation but a warm propagator is needed for successful results.*

Place the cuttings in a heated propagator with a temperature of 18–21°C (64.4–69.8°F). Alternatively, enclose the pot in a clear polythene bag supported with a few split canes and stand on a window sill in a warm room.

Young plants appearing at the base of the leaves are an indication that the cuttings have rooted. They can then be lifted and potted individually.

Sinningia and *streptocarpus* have large leaves which are a bit too long to be left whole. They can be cut in half and the bottom halves used as cuttings.

Small Sections of Leaves

The popular houseplant *Sansevieria trifasciata* (mother-in-law's tongue) has very long

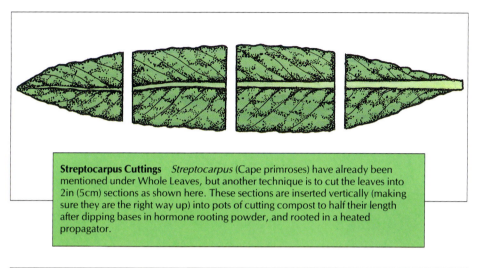

Streptocarpus Cuttings *Streptocarpus* (Cape primroses) have already been mentioned under Whole Leaves, but another technique is to cut the leaves into 2in (5cm) sections as shown here. These sections are inserted vertically (making sure they are the right way up) into pots of cutting compost to half their length after dipping bases in hormone rooting powder, and rooted in a heated propagator.

A mature leaf of Sansevieria trifasciata *can be cut into 2in (5cm) long sections to provide cuttings, which are rooted in a heated propagator.*

sword-like leaves. These can be used for propagation. However, do not propagate the yellow-edged cultivar 'Laurentii' from leaf cuttings as the resultant plants will have plain green leaves. Instead, this will have to be increased by division (*see* Greenhouse and Houseplants, page 7). Other species and cultivars of *sansevieria* can also be increased from leaf cuttings.

Take a mature leaf, lay it on a table and cut it into 2in (5cm) sections as shown here. It is important that they are kept upright as they must not be inserted upside-down. Dip the bases in hormone rooting powder then press them vertically into pots of cutting compost to half their length and root in a

The yellow-edged cultivar of Sansevieria trifasciata *should not be propagated from leaf cuttings but rather by division. As can be seen, it produces plenty of offsets.*

A leaf-bud cutting of clematis with one leaf removed, leaving two buds and a single leaf. Prepare them from soft shoots in the spring.

heated propagator (*see* Whole Leaves, page 33).

When rooted, a small plantlet will appear from the base of each cutting. They can then be potted individually.

Leaf-Bud Cuttings

The leaf-bud cutting is a variation on the leaf cutting and consists of a section of stem containing a leaf at the top with a dormant growth bud in the axil. Various hardy plants as well as tender kinds can be propagated from leaf-bud cuttings.

Leaf-bud cuttings of camellias should have a sliver of bark removed from the base to expose the wood. This increases the chances of rooting.

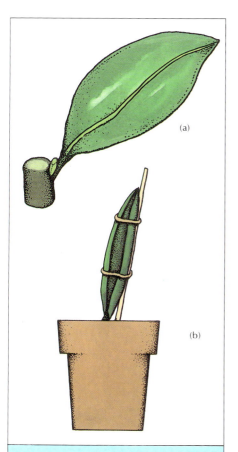

Ficus elastica *(rubber plant) is commonly propagated from leaf-bud cuttings but plenty of warmth is needed for successful results.*

Propagating Rubber Plants
Ficus elastica or rubber plant can be propagated from leaf-bud cuttings during spring or summer.
· **Preparing cuttings**   Choose a young shoot and cut it into sections consisting of a ⅔in (2cm) length of stem with one leaf and bud at the top (a). Dip the base in hormone rooting powder.
· **Insertion**   As the leaf is large, roll it lengthways and secure it with an elastic band, then insert the cutting in a 3in (8cm) pot of cutting compost, supporting it with a split cane as shown (b).
· **Rooting**   A heated propagator is needed with a bottom temperature of 21°C (69.8°F) and high humidity.

The tender plant most commonly propagated in this manner is *Ficus elastica* (rubber plant). Among the hardies, the climbing plants *clematis, passiflora* (passion flower) and *hedera* (ivy) are propagated by this method in the spring, using soft shoots. They are easy to prepare – simply cut some shoots into ⅖–⅘in (1–2cm) sections, each with a leaf at the top. Make the top cut immediately above a leaf and the basal cut up to ⅘in (2cm) below. If leaves are borne in pairs, remove one of the leaves.

Dip the bases in hormone rooting powder

then insert in pots of cutting compost. Only the leaf and bud should be visible above the compost. Water well in and root in a heated propagator (*see* Whole Leaves, page 33). Rooting is confirmed when the bud has made some growth.

Camellias can be propagated in the same way during late summer, using semi-ripe current year's shoots (*see* Semi-Ripe Cuttings, page 23). Remove a sliver of bark from the base of each cutting to expose the wood. Camellia cuttings are not easy to root and need a basal temperature of 21°C (69.8°F) and high humidity. They take about two months to form a good root system.

Leaf-bud cuttings of blackberries and loganberries are easily rooted in a garden frame during late summer.

Miscellaneous

There are several other unusual types of cuttings, including eye cuttings for grape vines, lily scales, pipings for pinks and ready-rooted (or Irishman's) cuttings.

Lily Scales

If you inspect a lily bulb, you will see that it is composed of fleshy scales. These can be used for propagation in late summer or spring. Dig up a bulb and carefully snap off a few scales, then replant.

Insert the scales upright to half their length in a container of cutting compost (equal parts peat and sand). Root them in a heated propagator, on the greenhouse bench or in a garden frame. Water sparingly.

Each scale produces a small bulb at its base when rooted. Then the scales can be potted individually, using potting compost. Once established in the pots, harden the bulbs in a garden frame if necessary then plant in a nursery bed outdoors. The young bulbs will take two to three years to produce flowers.

Pipings

Border and rock-garden pinks are propagated from pipings (*see also* Layering Carnations, page 15). Pipings are a type of cutting obtained by snapping out the tops of young shoots. They will break off just below

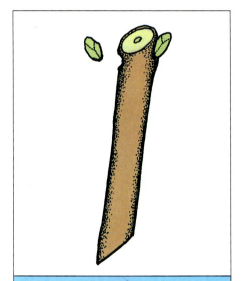

Eye Cuttings of Grape Vine
Eye cuttings are used for the fruiting *Vitis vinifera* cultivars and for the ornamental vines like 'Brant', *V. coignetiae* and *V. davidii*.
• **Preparation** Cuttings are prepared with secateurs from well-ripened, woody 1-year-old stems during early to mid-winter, making them into 1–1⅓in (2.5–3.5cm) sections, removing one of the buds to leave just one, as shown here. Dip bases in hormone rooting powder.
• **Insertion and rooting** Push cuttings down into pots of cutting compost so that only the buds are showing. Water in and place in a propagator with a bottom temperature of 21°C (69.8°F). Top growth will occur before rooting takes place.

Lilies grow from bulbs and the fleshy scales of these can be used for propagation, either in late summer or spring. They root easily with or without artificial heat.

Lily bulbs are composed of fleshy scales, (left) some of which can be snapped off and encouraged to root, when they will form new bulbs. Insert lily scales upright to half their length in a container of cutting compost, (right) and root with or without artificial heat.

Pipings of border pinks, obtained by snapping out the tops of young shoots. Each must have at least three pairs of leaves.

Pipings, a type of cutting, are used to propagate border pinks during the period mid- to late summer. They can be rooted in a garden frame.

a node (or leaf joint). Choose sturdy shoots and take them during the period mid- to late summer.

Each piping should have three to four pairs of leaves when the lower pair has been pulled off, but these will vary in length depending on the type of pink. The shortest ones will be those of rock-garden pinks.

Insert pipings up to their lower leaves around the edge of a pot filled with cutting compost and then water in. Root them in a garden frame. They will produce substantial roots within six weeks, when the young plants can either be potted individually in 3½in (9cm) containers and wintered in the frame, or planted direct in the garden.

It is a good idea to propagate pinks regularly as they have only a short life. Plants start to deteriorate after three or four years when they should be replaced by young specimens.

Ready-Rooted Cuttings

These are popularly known as Irishman's cuttings and consist of rooted shoots which are carefully pulled away from the parent plant. These are then potted into small containers and placed in a garden frame to grow to a suitable size for planting out in the garden.

The method can only be used for plants that produce new shoots from their base or root area, such as chrysanthemums and asters (Michaelmas daisies).

The young shoots should be removed in the growing season, in the spring. However, pansies (*viola*) are propagated by this method in late summer. With these, cut back the top growth by about one-half and then place some potting compost around the remainder of the stems. The stems will root into this and can then be cut off, complete with roots. Pot them and winter in a garden frame.

A ready-rooted pansy (viola) cutting, obtained by mounding potting compost around the bases of cut-back stems in late summer.

Propagation from leaf cuttings, described earlier in this chapter, often results in many plantlets, as with these streptocarpus or Cape primrose cuttings.

6 • OUTDOOR SEED SOWING

Seeds can be saved from one's own plants or they can be bought from seedsmen. A combination of the two is perhaps a sensible option as it is not advisable to save seeds from hybrid plants or cultivars because the resultant seedling will not be the same as their parents. It is always advisable to save seeds only of species, the offspring of which will be identical to the parent plants.

Seed pods and capsules turn from green to brown as the seeds ripen. These poppy seed heads are not yet ready for harvesting.

Collect seeds of hardy plants when ripe, indicated in fruits and berries (such as sorbus *or mountain ash) by a change of colour.*

Collecting Seeds

Collect seeds of hardy plants like shrubs, perennials, annuals, bulbs and alpines as soon as they are ripe, which is in late summer and autumn for the majority. Signs of ripeness are fruits and berries turning colour and seed pods and capsules changing from green to brown.

Seed pods and capsules, but not fruits and berries, should be dried after collection by laying them on sheets of newspaper for a few weeks in a sunny, dry and airy place.

Seed pods and capsules should be dried after collection by laying them on paper for a few weeks in a sunny, dry airy place, when they will split open.

The seed containers will split open and you can then extract the seeds. If they do not split, gently crush them. Separate the seeds from the debris and store them in envelopes, in a cool dry place for the winter. Remember to label the envelopes.

Preparation of Seed Beds

The site for seed sowing should be prepared thoroughly. Start by digging in autumn and leaving the ground rough over winter for the frost to work on it. Then, as soon as the soil is in a fit state to work during the spring, when it is drying out and warming up, you can prepare seed beds as required.

Break down the roughly dug ground with a fork, level it, then firm it by treading over

Final preparation of an outdoor seed bed consists of raking with an iron rake until the soil is broken down finely and is level.

the site systematically, with the weight on your heels. Then you can apply a general-purpose fertilizer and carry out final preparation by raking. Using an iron rake, work in various directions until the soil is broken down finely and the top ⅖–1in (1–2.5cm) is loose. This is the ideal state for seed sowing.

Sowing in Open Ground

Seeds of most hardy plants, including vegetables, can be sown in the open ground during spring. However, very fine or dust-like seeds (such as those of rhododendrons) would be better sown in containers and germinated under glass (*see* Chapter 7).

Hardy perennials and biennials are sown in a well-prepared seed bed during late spring or early summer. The seedlings are then transplanted to nursery beds to grow on and are planted in their final positions during autumn.

Hardy annuals should be sown direct in their flowering positions during early or mid-spring. Choose a warm sunny spot with well-drained soil. They are most effective if sown in bold informal groups, but sow the seeds in drills across the groups, spaced 6in (15cm) apart.

If the soil is very well drained, tough hardy annuals can be sown in early autumn, when they will flower earlier the following year. They should be covered with cloches over winter.

For the timing of vegetable sowings you must be guided by the information on the seed packets or in the catalogues.

Methods of Sowing

The easiest way to sow seeds outdoors is in

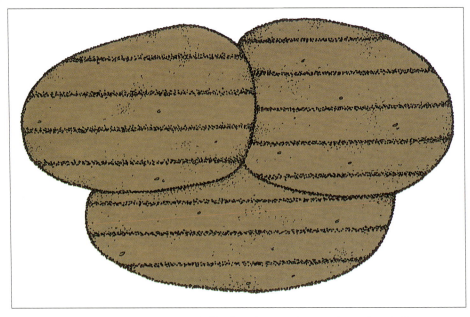

Hardy annuals are most effective if sown in bold informal groups, but sow the seeds in shallow drills across the groups, spaced 6in (15cm) apart.

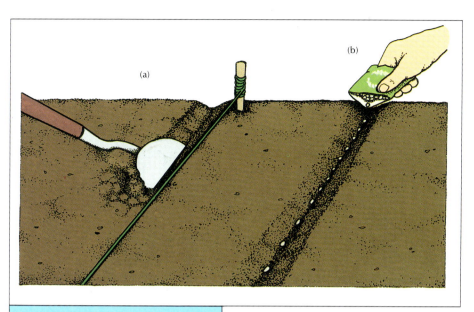

(a)
(b)

Sowing in Drills

A drill is simply a shallow furrow in which to sow seeds. For the majority of small seeds this can be about ⅖in (1cm) deep, but be guided by the information on seed packets or in catalogues.

· **Making drills**   First put down a garden line where you want the drill, then run along it with a draw hoe, the corner of which takes out the furrow as shown here (a). Alternatively, use a pointed stick.

· **Sowing**   Sow either direct from the packet as shown (b) or take a pinch of seeds between finger and thumb and slowly release the seeds into the drill. To sow thinly, move along the drill quite quickly but release the seeds slowly. After sowing, cover the seeds by gently raking fine soil over them, making sure that the depth is even.

needed, such as for early carrots and turnips or spring onions. Remember that weeding will be more difficult. Broadcast sowing is always used for sowing grass seed to create a new lawn. The technique is to take a quantity of seed in a hand and to sow with a side-to-side sweeping action about 12in (30cm) above the ground, slowly releasing the seeds. The seeds are covered by gently raking them into the soil.

Seeds must always be sown as thinly as possible to avoid overcrowding and to minimize thinning out of seedlings. No seeds should be sown closer than ⅖in (1cm) apart.

Special Technique for Woody Plants

Seeds of trees, shrubs and conifers can be sown in a raised bed during the spring. This is a good idea if the ground is inclined to lie wet over winter.

Throw up the soil to make a 3ft (1m) wide bed about 3–4in (8–10cm) high. Sow in

rows, placing them in drills. This is explained in the accompanying panel. However, seeds are sometimes scattered evenly over the site and this is known as 'broadcast sowing'. It is best used only where minimum thinning of seedlings is

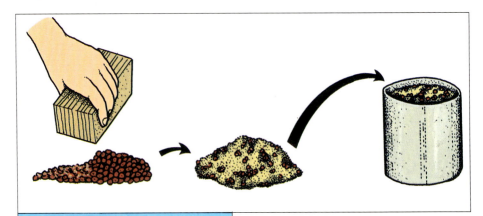

Berries and Fruits

The fleshy berries and fruits of shrubs and trees (e.g. rose, holly, *cotoneaster*, *berberis* and crab-apples) are not stored by the normal method (*see* Collecting Seeds, page 42) but by a technique known as stratification. This guarantees better germination.

• **Preparation** The berries are gently crushed to expose the seeds, then mixed with moist sand.

• **Storage** Place the mixture in a pot or tin can and store in a north-facing position outdoors – the colder the spot, the better.

• **Sowing** Sow the seed and sand mixture in spring (*see* Special Technique for Woody plants, page 45). Some may not germinate until the second spring.

Many shrubs and other plants produce their seeds in capsules which should be collected just before they split open, as described at the beginning of this chapter.

⅕–⅖in (0.5–1cm) deep drills across the bed, spaced 4in (10cm) apart, and cover the seeds by spreading a thin layer of pea shingle over the entire bed.

Soil can be used for covering seeds, but pea shingle is better as it stops birds from disturbing the ground, prevents weed growth and helps the soil retain moisture.

The seedlings can be left in the bed until the autumn or following spring, when they should be lifted and planted elsewhere. Some seeds may not germinate until the second spring.

7 • SOWING SEEDS UNDER COVER

Greenhouse plants, house-plants, pot plants and summer bedding plants are raised in a heated greenhouse, or, failing that, on a window sill indoors. Tender vegetables are raised in the same conditions, as well as hardy kinds intended for very early crops. The majority of these are sown in spring, summer bedding being raised in the period late winter to mid-spring, depending on speed of growth.

Requirements

You will need a heated greenhouse, ideally with a minimum temperature of 7–10°C (44.6–50°F). (Electric fan heaters are very efficient and reliable.) An electrically heated propagator is also virtually essential,

to provide bottom heat for seed germination. It must be fitted with a thermostat. A good germination temperature for most seeds is 18–21°C (64.4–69.8°F). Most people sow in soil-less compost or multi-purpose compost these days; and you will need some plastic labels.

As mentioned earlier, seed raising can be carried out on a window sill in a warm room indoors, and there are available small electric window sill propagators which are very economical to run.

Preparing Containers

To prepare a container for sowing, using soil-less compost, fill it to overflowing and scrape off the surplus with a piece of wood.

With a small, heated greenhouse you will be able to raise many plants from seeds, such as summer bedding plants, pot plants and tender vegetables.

An electrically heated propagator to provide bottom heat, whether homemade (as here) or proprietary, is virtually essential for successful seed germination under glass.

Then, using a wooden presser that exactly fits the container, firm the compost lightly. This will result in watering space at the top and a smooth level surface on which to sow.

Sowing Techniques

Seeds are sown broadcast (or scattered) over the surface of the compost as evenly as possible. Most people make the mistake of sowing too thickly. Seeds must on no account be touching and should be a minimum of ⅕in (6mm) apart each way. Large seeds can be spaced out individually.

Very tiny dust-like seeds are the most difficult to sow thinly and evenly. Take a pinch of seeds for the container (as a guide, a thin layer spread over a one pence piece will be enough for a standard-sized seed

Seed raising can be carried out on a window sill and there are various small propagators for the purpose, from the simplest non-heated covered seed tray (as shown) to small electric models.

Containers for Seed Sowing A great depth of compost is not needed for seed raising so choose the shallowest containers you can find. Small quantities of seeds can be sown in 4in (10cm) half pots; larger amounts in full- or half-size seed trays. Containers for seed sowing must be kept scrupulously clean so wash used ones thoroughly.

The easiest method of covering seeds is to sift the compost over them, using a horticultural sieve with a fine or medium mesh size.

After sowing seeds, moisten the compost by standing the containers almost up to their rims in water. Remove when the compost surface becomes wet.

tray) and mix it with some very fine dry silver sand to make handling easier.

The actual techniques of sowing tiny and medium-sized seeds are shown in the accompanying panel. All except very tiny seeds, such as those of *begonia*, should be covered with a layer of compost, the depth equalling approximately twice the diameter of the seeds. The easiest method of covering seeds is to sift the compost over them, using a sieve with a fine or medium mesh size. The layer can then be very lightly firmed with your wooden presser.

After sowing, moisten the compost by standing the container almost up to its rim

Seed Sowing
There are two techniques for even and thin sowing of tiny to medium-sized seeds. One is to hold a quantity of seeds in the palm of one hand, raise it 6in (15cm) above the compost surface and then move it to and fro, at the same time tapping it with your other hand to slowly release the seeds (a).
• **From paper** Alternatively, the seeds can be sown from a sheet of folded paper (b).
• **Even sowing** To ensure really even sowing, scatter half the quantity of seeds in one direction and the other half at right angles. Move your hand or the paper fairly rapidly over the compost surface, but release the seeds slowly.

(a)

(b)

in water until the surface of the compost becomes moist, then remove and allow to drain. Add the fungicide, Cheshunt Compound, to the water to prevent the fatal seedling disease, damping off.

Then place the containers in the propagator. To prevent the compost drying out, which would inhibit germination, cover them with sheets of black polythene or newspaper. These must be removed as soon as germination occurs as seedlings need to have really good light, but shade them from direct sunlight at all times or they could get scorched.

Pricking Out

This is an odd gardening term. It simply means transplanting seedlings into larger containers (usually deep seed trays), as soon as they are large enough to handle easily, to give them more room to grow.

Forty to forty-five seedlings can be planted in rows in a standard-size seed tray. However, large-growing plants like bedding pelargoniums and dwarf bedding dahlias would be better pricked out into individual 3½in (9cm) pots.

Soil-less potting or multi-purpose compost is generally used for pricking out these days. The technique is shown in the accompanying panel (page 52).

The seedlings should be grown on in a heated greenhouse, or on your lightest window sill in a warm room indoors, shading them from sunshine and checking that the compost does not dry out.

After pricking out, seedlings can be grown on in a heated greenhouse, ensuring they receive maximum light but shading them from sunshine.

Transplanting Seedlings

The technique of transplanting, or pricking out, involves lifting a few seedlings at a time with a dibber (a pencil is a good substitute). Seedlings should be handled only by the seed leaves (a), never by the stems. Tiny seedlings can be held with a small forked stick.

· **Planting**   Use the dibber to make a hole for each seedling sufficiently deep and large to enable the roots to dangle straight down (b). The seed leaves should be just above the compost surface. Then gently push some compost around the seedling with the dibber and lightly firm it.

· **Watering in**   After pricking out, water in seedlings with a solution of Cheshunt Compound applied with a watering-can fitted with a fine sprinkler.

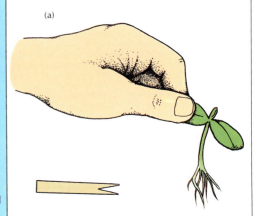

(a)

(b)

Summer bedding plants, like impatiens *or busy lizzie, should be hardened off in a garden frame prior to planting out, when danger of frost is over, by which time they should be coming into flower.*

Any plants intended for planting oudoors should be hardened in a garden frame for at least two or three weeks prior to planting out. Remember that tender plants such as summer bedding and vegetables like tomatoes should not be planted out until all danger of frost is over, i.e. late spring or early summer.

There is a tried and tested procedure for hardening plants, or gradually acclimatising them to outdoor conditions.

For the first few days after placing the plants in a frame, keep the frame covers or 'lights' closed. Then open the 'lights' a little by day but close them again at night. Then over a period of two or three weeks, gradually open the 'lights' more and more by day but still close them at night. Towards the end of this period they are removed completely by day but closed again at night.

A few days before you intend setting the plants out in the garden, the frame 'lights' should be left off at night also. By now the plants will be fully hardened or acclimatised.

8 • BUDDING

Budding is quite a skilled method of plant propagation but it is still within the capabilities of home gardeners. It can be used for a number of different plants, including ornamental and fruiting trees, but rose budding is perhaps the most popular for the amateur.

Put very simply, budding involves inserting a growth bud of the cultivar you wish to propagate into another plant which provides the root system, known as the rootstock. All being well, the bud eventually grows, producing shoots and leaves which become the top part of the new plant.

Why go to all this trouble? Why not raise roses from cuttings? Most roses can indeed be propagated from cuttings, but few grow well on their own roots, resulting in weak, disappointing plants. They need a much more vigorous root system, hence the use of a rootstock, which should be a strong wild rose such as *Rosa canina* (dog rose).

Obtaining and Planting Rootstocks

You will have to raise rootstocks from seeds

A bud is obtained by inserting a knife at least ⅖in (1cm) below it, drawing it underneath and bringing it out a similar distance beyond the bud.

(*see* Special Technique for Woody Plants, page 45). Alternatively, it is sometimes possible to buy small quantities from mail-order nurserymen. Look through the classified advertisements in the gardening press.

One-year-old seedling rootstocks are planted in a nursery bed during mid- to late autumn. Set them 12in (30cm) apart in a row. Rows should be spaced 2¾ft (90cm) apart to give you room to work.

There is a special planting technique for rose rootstocks. The stem (known as the neck), between the roots and the shoots, must be kept above ground level. After planting, soil is mounded up around it. It should be kept moist, and watering may be required during dry periods in the following spring and summer.

Preparing for Budding

The rootstocks are budded in the summer following planting. Just before inserting buds, the soil around the rootstock stems must be pulled away, taking care not to damage the bark. The stems should be wiped with a piece of soft cloth to remove any soil.

Growth buds of the cultivar you wish to propagate should be on current year's shoots and they must be firm and ripe. Remove complete shoots and cut off the soft tops. Then cut off all of the leaves, but leave the leaf stalks intact. These shoots containing the buds, popularly known as 'bud sticks', are best stood in a bucket of water until the buds are removed. Then all is ready for budding.

The Technique

You will need a really sharp horticultural knife for budding – ideally a proper budding knife (also useful for taking cuttings). The rootstock cut is made and a bud immediately

It is not only commercial rose growers who are able to propagate roses from budding: it is perfectly within the capabilities of home gardeners, albeit on a smaller scale. Often, the bushes start flowering in their second summer from budding.

inserted. Remove only one bud at a time from the bud stick.

A bud is removed on a shield-shaped piece of bark and is obtained by inserting the knife at least ⅖in (1cm) below a bud, drawing the knife underneath it and bringing it out again the same distance beyond.

If done accurately, there will be a thin

Inserting a Bud

- **The rootstock cut** Make a T-shaped cut in the neck of the rootstock, in the bark, as close as possible to the ground. The stem of the 'T' should be 1–1½in (2.5–4cm) in length and the bark on each side carefully lifted (a).
- **Inserting the bud** Holding the bud by the leaf stalk, gently push the shield-shaped piece of bark down behind the bark of the rootstock (b). Any surplus bark protruding above the rootstock cut should be cut off.
- **Securing the bud** The bud can be tied in with a few firm twists of raffia, but do not actually cover it. Alternatively, use a modern thin rubber budding patch which does, in fact, cover the bud. Place the patch over the bud, stretch the two ends to the back of the rootstock and secure them by means of the wire staple-like clip.

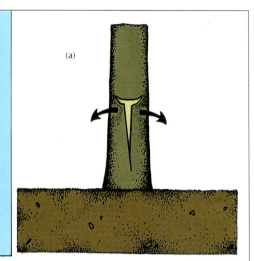

(a)

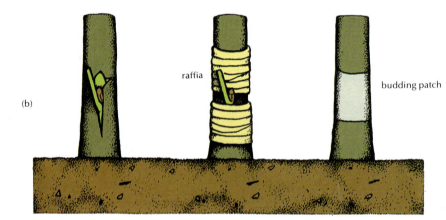

(b)

raffia

budding patch

slice of wood at the back of the 'shield' and this should be carefully removed by prizing it off with the blade of the knife (or with the spatula on the handle in the case of a proper budding knife).

Subsequent Operations

After inserting the buds, there is little to be done until the following late winter or early spring, when the top of the rootstock is cut off, with secateurs, just above the bud. However, do continue to water during the first summer if the soil starts to dry out. During the second summer, feed and water the young roses, some of which may produce a few flowers. In the second autumn, the plants can be lifted and planted in their final positions.

9 • GRAFTING

Grafting is an even more skilled method of propagation than budding. Nevertheless, keen amateur gardeners do often have a go – and achieve success. It is carried out for the same reasons as budding – to provide plants with a rootstock to ensure that they grow well. With fruit trees, rootstocks are used to control the vigour of the trees. Grafting also enables one to increase plants that may be difficult from cuttings. It is only used for cultivars, not species.

Grafting differs from budding in that a piece of stem from the cultivar to be propagated (known as the scion) is united with a rootstock to produce the top growth of the plant.

Young trees propagated by whip and tongue grafting. It will take four years to produce a sizeable tree for planting in its final position.

Whip-and-Tongue Grafting

This graft is used to propagate ornamental and fruiting trees and is one of the easier techniques. It is important to obtain the correct rootstocks. (The grafting of fruit trees like apples and pears will not be included here as home gardeners cannot obtain suitable rootstocks.) Instead, stick to ornamental trees, choosing the common counterpart for the rootstock, which can be seed-raised. (*See* Chapter 6, Special Techniques for Woody Plants, page 45.)

Rootstocks

Here are some examples: *crataegus* cultivars (thorns) on *C. monogyna* (hawthorn) rootstocks; *laburnum* cultivars on *L. anagyroides* (common laburnum) rootstocks; *malus* (ornamental crab-apples) on *M. sylvestris* (common crab-apple) rootstocks; *prunus* (ornamental cherries) on *P. avium* (wild cherry) rootstocks; and *Sorbus aucuparia* (mountain ash) cultivars on *S. aucuparia* rootstocks.

Rootstocks should be 2 years old and lined out in a nursery bed one year before grafting. Plant in autumn, spacing them 2ft (60cm) apart in a row.

The Procedure

Grafting takes place during early spring and the technique is illustrated in the accompanying panel. It is essential to use a very sharp horticultural knife – ideally a proper grafting knife – and to make really clean and smooth cuts.

The cuts must match in size and be perfectly flat or they may not unite. The layer of green tissue just under the bark of scion and the rootstock (exposed when the cuts are made) must be in close contact. If you cannot get an exact match all round a graft, then at least check that the layers match on one side.

If the graft is successful, the scion will

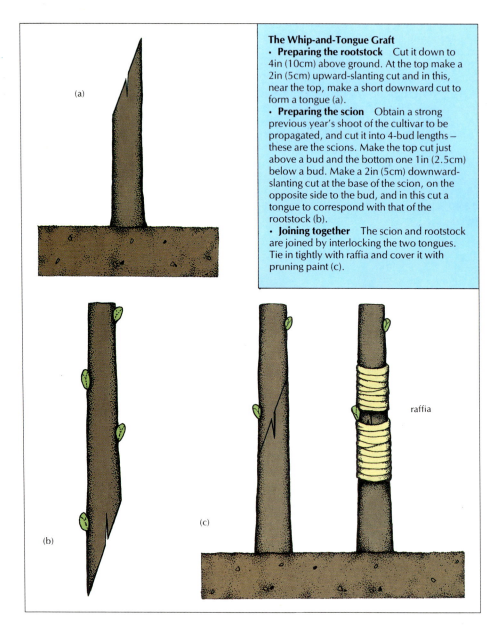

(a)

The Whip-and-Tongue Graft
· **Preparing the rootstock**   Cut it down to 4in (10cm) above ground. At the top make a 2in (5cm) upward-slanting cut and in this, near the top, make a short downward cut to form a tongue (a).
· **Preparing the scion**   Obtain a strong previous year's shoot of the cultivar to be propagated, and cut it into 4-bud lengths – these are the scions. Make the top cut just above a bud and the bottom one 1in (2.5cm) below a bud. Make a 2in (5cm) downward-slanting cut at the base of the scion, on the opposite side to the bud, and in this cut a tongue to correspond with that of the rootstock (b).
· **Joining together**   The scion and rootstock are joined by interlocking the two tongues. Tie in tightly with raffia and cover it with pruning paint (c).

raffia

(c)

(b)

start growing later in the spring, at which time the raffia holding it in place should be slit. It will take four years to produce a size-able tree for planting in its final position. To obtain a straight trunk, tie it in regularly as it grows to a stout bamboo cane.

(a)　　　　　　　　　　　　　　　(c)

raffia

Saddle Grafting

This graft is used for hybrid rhododendrons. These shrubs are not easy to propagate by other means. Rootstocks can be 2- or 3-year-old plants of *R. ponticum* (the common purple rhododendron). They should be raised from seeds in a cool greenhouse (*see* pages 47–53) and grown in pots outdoors. Alternatively, use rootstocks of *R.* 'Cunningham's White', which is quite easily raised from semi-ripe cuttings in heat (*see* Semi-Ripe Cuttings, page 23).

Saddle grafting is carried out in early spring in a heated greenhouse. The rootstocks should be moved under glass two to three weeks before you want to graft. The scions are prepared from previous year's shoots, each with a terminal growth bud. Scions and rootstocks should be of the same thickness.

The technique of saddle grafting is shown in the accompanying panel. Place the grafts in a heated propagating case with a minimum temperature of 18°C (64.4°F) and high humidity. They can be laid on their sides if there is insufficient headroom, with the grafts facing upwards. The grafts will unite within about six weeks and when the union looks strong the raffia can be cut away and the plants put on the greenhouse bench to acclimatize to cooler conditions. Complete the hardening process in a garden frame prior to planting in the garden. Grafted plants will take two to three years to attain flowering size.

SEASONAL TASKS

Early Spring	Divide hardy herbaceous and evergreen perennial plants. Divide clump-forming and carpeting alpines. Divide greenhouse plants and house-plants. Divide congested clumps of *galanthus* (snowdrops). Take soft basal cuttings of hardy and tender perennials. Increase lilies from scales. Take Irishman's cuttings when available. Final preparation of seed beds. Sow seeds outdoors and under cover. Sow seeds of hardy annuals outdoors. Sow seeds of trees, shrubs and conifers outdoors. Sow seeds of summer bedding plants under cover. Prick out seedlings under cover. Cut off tops of budded rose rootstocks. Whip-and-tongue grafting of ornamental trees. Saddle grafting of rhododendrons.
Mid-Spring	Divide hardy herbaceous and evergreen perennial plants. Divide aquatics, including water lilies. Divide greenhouse plants and house-plants. Plant gladioli corms and cormlets outdoors. Divide large clumps of dahlias and plant in garden. Simple layering of shrubs until late summer. Air layering of trees and shrubs until late summer. Serpentine layering of climbers until late summer. Leaf cuttings of tender and hardy plants until late summer. Leaf-bud cuttings of *Ficus elastica* (rubber plant) and throughout summer. Leaf-bud cuttings of climbers: *clematis, passiflora* (passion flower) and *hedera* (ivy). Increase lilies from scales. Take Irishman's cuttings when available. Final preparation of seed beds. Sow seeds outdoors and under cover. Sow seeds of hardy annuals outdoors. Sow seeds of trees, shrubs and conifers outdoors. Sow seeds of summer bedding plants under cover. Prick out seedlings under cover.
Late-Spring	Divide aquatics including water lilies. Divide clumps of bulbs and corms when dormant and remove bulblets and cormlets. Softwood cuttings of hardy and tender plants. Soft basal cuttings of hardy and tender perennials. Leaf-bud cuttings of climbers: *clematis, passiflora* (passion flower) and *hedera* (ivy). Take Irishman's cuttings when available. Final preparation of seed beds. Sow seeds outdoors and under cover. Sow seeds of hardy perennials and biennials outdoors. Prick out seedlings under cover. Plant out tender plants like summer bedding and vegetables, such as tomatoes.
Early Summer	Divide aquatics, including water lilies. Air layer house-plants and continue through summer. Propagate from plantlets and continue through summer. Propagate strawberries from runners. Softwood cuttings of hardy and tender plants. Sow seeds of hardy perennials and biennials outdoors. Plant out tender plants like summer bedding and vegetables, such as tomatoes. Carry out rose budding. Prick out seedlings under cover.

Mid-Summer	Tip layering of blackberries, loganberries and similar until late summer. Propagate strawberries from runners. Semi-ripe cuttings of shrubs, conifers, half-hardy perennials, greenhouse plants and house-plants. Take pipings of border and rock-garden pinks.
Late Summer	Last chance for tip layering of blackberries, loganberries and similar. Last chance for simple, air and serpentine layering. Layering of border carnations and pinks. Plant young strawberries. Semi-ripe cuttings of shrubs, conifers, half-hardy perennials, greenhouse plants and house-plants. Last chance for taking leaf cuttings of tender and hardy plants. Take leaf-bud cuttings of camellias, blackberries and loganberries. Last chance to take leaf-bud cuttings of *Ficus elastica* (rubber plant). Propagate lilies from scales. Take pipings of border and rock-garden pinks. Take Irishman's cuttings of pansies (*viola*). Seed collecting through to late autumn. Carry out rose budding.
Early Autumn	Plant hardy bulbs outdoors. Plant young strawberries. Sow tough hardy annuals if soil is very well drained and cover with cloches over winter. Stratification or winter storage of fleshy fruits and berries.
Mid-Autumn	Divide hardy herbaceous and evergreen perennial plants if soil is very well drained. Plant hardy bulbs outdoors. Lift gladioli, remove cormlets and store both over winter in a cool frost-proof place. Lift and store dahlias for the winter in a cool frost-proof place. Semi-ripe cuttings of shrubs and conifers. Dig sites for seed beds. Stratification or winter storage of fleshy fruits and berries. Plant rose rootstocks.
Late Autumn	Divide hardy herbaceous and evergreen perennial plants if soil is very well drained. Propagate shrubs from suckers between now and early spring. Hardwood cuttings of shrubs and soft fruits. Last chance for seed collecting. Dig sites for seed beds. Stratification or winter storage of fleshy fruits and berries. Plant rose rootstocks.
Early Winter	Take root cuttings of hardy shrubs and perennials. Take eye cuttings of grape and ornamental vines.
Mid-Winter	Take root cuttings of hardy shrubs and perennials. Take eye cuttings of grape and ornamental vines.
Late Winter	Start sowing summer bedding plants under cover. A good period to stock up with seed composts, plant labels, pots and seed trays. Cut off the tops of budded rose rootstocks. Move *rhododendron* rootstocks into a heated greenhouse in readiness for saddle grafting during early spring.

GLOSSARY

Alpine Plant originating from mountainous areas. The term is loosely used to include any small plants suited to growing on rock gardens.
Annual A plant that grows to maturity, flowers, sets seeds and then dies, within the space of one year.
Aquatic A plant that lives in water.
Axil The angle where a leaf joins a stem.

Bedding plant A plant used for temporary garden display, mainly in spring or summer.
Biennial A plant that grows to maturity, flowers, sets seeds and dies over two growing seasons.
Bud A dormant compacted shoot containing embryo leaves and/or flowers.
Budding A method of propagation whereby a dormant growth bud is inserted into the stem of a rootstock. The bud eventually forms the top growth; the rootstock, the roots of the new plant.
Bulb An underground storage organ. It stores water and food, and contains an embryo flower.
Bulblet An offset produced at the side of a parent bulb.

Climber A long-stemmed plant that supports itself by various means.
Cutting compost A compost specially prepared for rooting cuttings, often composed of equal parts of peat and sand, with no fertilizers added.
Conifer A tree or shrub that produces cones, which contain seeds.
Corm A storage organ similar to a bulb, containing food, water and an embryo flower.
Cormlet An offset produced alongside the parent corm.
Crown The ground-level part of a perennial plant which produces shoots and roots.
Cultivar A variant of a plant that originated under cultivation.
Cuttings Shoots, stems, leaves or roots which are removed from plants and encourage to form roots of their own, when they develop into new plants.

Deciduous This term refers to a plant that sheds its leaves in the autumn, e.g. a tree, shrub or climber.
Dibber A stick with a blunt-pointed end, rather like a fat pencil, used when pricking out seedlings.
Division The technique of splitting a plant into a number of small pieces, each containing roots and buds or stems.
Dormant A resting period, when a plant achieves little or no growth.
Drill A straight shallow furrow made in the soil in which to sow seeds.

Evergreen This term refers to a plant that retains leaves all the year round, e.g. a shrub, tree or conifer.

Fertilizer Material that supplies plant foods.

Genus This is a botanical classification and is the first name of a plant, e.g. *Rosa* (rose). A genus contains one or more species. The plural form is genera.
Germination The process whereby a seed starts to grow, or develop into a seedling.
Grafting A method of propagation whereby a piece of stem, known as the scion, is united with the stem of a rootstock. The scion eventually forms the top growth; the rootstock, the roots of the new plant.

Half-hardy Any plant unable to survive a winter outdoors, but not requiring protection all year round.
Harden To gradually acclimatize plants to cooler outdoor conditions, especially those raised in a warm place.
Hardy Any plant that will survive winters outdoors without any form of protection.
Heel A small piece of older tissue or wood attached to the base of a cutting.

Herbaceous A perennial plant whose stems die down for the winter.
Hormone rooting powder A special preparation which induces root formation in cuttings.
Humidity Amount of moisture in the air.
Hybrid A plant which originates from the crossing of two species, cultivars or varieties.

Lateral A shoot which develops on a main stem – a side shoot.
Layering Encouraging a stem to produce roots while it is still attached to the parent plant.

Node The part of a stem from which one or more leaves are produced.

Offset A new plant produced alongside the parent, which can be removed and planted elsewhere.

Perennial A plant that lives for more than three growing seasons. The term normally refers to non-woody plants, which may be herbaceous or evergreen.
Piping A cutting of a border carnation or pink.
Potting and seed compost A mixture of various materials, organic and inorganic, including fertilizer, which is used to grow plants in containers.
Pricking out Transplanting seedlings raised under cover to give them more room to grow.

Rhizome An underground stem that creeps through the soil, spreading the plant, and which stores food and water.
Rootstock The part of a plant that forms the root system when propagating plants by budding or grafting.
Runner A stem that grows over the surface of the soil, producing new plants. Strawberry plants produce runners.

Scion A piece of stem containing growth buds from a plant to be propagated by grafting. It is united with the stem of a rootstock.
Seed The unit of reproduction of a flowering plant.
Seed bed Ground specially prepared for seed sowing.
Seed leaves The first leaves produced by seedlings.
Seedling A plant which has recently germinated or emerged from a seed.
Shrub A plant with a number of woody stems, living for an indefinite period.
Species A plant which is contained within a genus. The specific name is the second name of a plant, e.g. *Rosa canina*, *canina* being the species. Species are wild plants but many are grown in gardens.
Stratification This is a method of storing fleshy fruits and berries over winter prior to spring sowing, subjecting them to alternate freezing and thawing.
Sucker A shoot which grows from a subterranean stem or root.

Tender Any plant that would be damaged by frost.
Thinning Removing surplus seedlings, so giving the remainder more room to grow.
Tilth The surface layer of soil when it has been cultivated to a fine crumbly state. Ideal conditions for seed sowing, for example.
Tree A woody plant with a single branchless stem or trunk supporting a head of branches at the top, which is known as the crown.
Tuber A swollen subterranean root or stem that contains plant foods and water, so acting as a storage organ for the plant. The dahlia is a well-known tuber-producing plant.

Variety A variant of a species, which occurs only in the wild. People often use the term to refer to cultivars, but this is an incorrect usage.

INDEX